W9-ALM-080

SPACE BUSTERS

SPACE TRAVEL

Stuart Atkinson

A Harcourt Company

Austin New York

www.raintreesteckvaughn.com

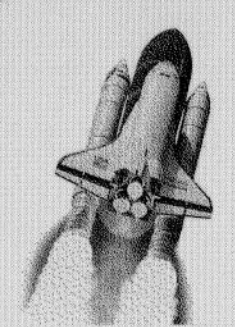

LOOK FOR THE SHUTTLE

Look out for boxes like this with a space shuttle in the corner. They contain extra information and amazing space-buster facts and figures.

Published by Raintree Steck-Vaughn Publishers, an imprint of Steck-Vaughn Company

Designer: Tim Mayer
Editors: Jason Hook, Pam Wells
Consultant: Steve Parker

Library of Congress
Cataloging-in-Publication Data
Atkinson, Stuart.
Space travel / Stuart Atkinson.
p. cm.—(Space busters)
Includes index.
Summary: Provides an overview of efforts to learn more about our universe, describing space probes, astronauts' training and work, space shuttles and space stations, and discoveries about Mars.

ISBN 0-7398-4852-6
1. Manned space flight—Juvenile literature.
2. Interplanetary voyages—Juvenile literature.
[1. Manned space flight. 2. Outer space—Exploration.] I. Title. II. Series.

TL793.A885 2002
629.45—dc21 2001041728

Printed in Hong Kong.
Bound in the United States.
1 2 3 4 5 6 7 8 9 0 LB 05 04 03 02 01

Acknowledgments
We wish to thank the following individuals and organizations for their help and assistance and for supplying material in their collections: Corbis 9 bottom, 11 top (James A Sugar), 17 (Reuters NewMedia Inc); MPM Images 4, 5 top, 7 top, 8, 14, 18 top, 24 bottom, 28, 29, 30, 31; NASA 2, 12 bottom, 13, 20 bottom, 21, 23; Science and Society Picture Library 6 (NASA), 11 bottom (NASA); Science Photo Library 1 (Chris Butler), 3 (Julian Baum), 7 bottom (Chris Butler), 9 top (NASA), 10 (NASA), 12 top (Novosti), 16 top (NASA), 16 bottom (Victor Habbick Visions), 18 bottom (Julian Baum), 19 (US Dept of Energy), 20 top (Claus Luna/ Bonnier Publications), 22 (Julian Baum), 24 top (Chris Butler), 25 (Julian Baum), 26 (Novosti), 27 (Julian Baum); Topham Picturepoint 5 bottom, 15.

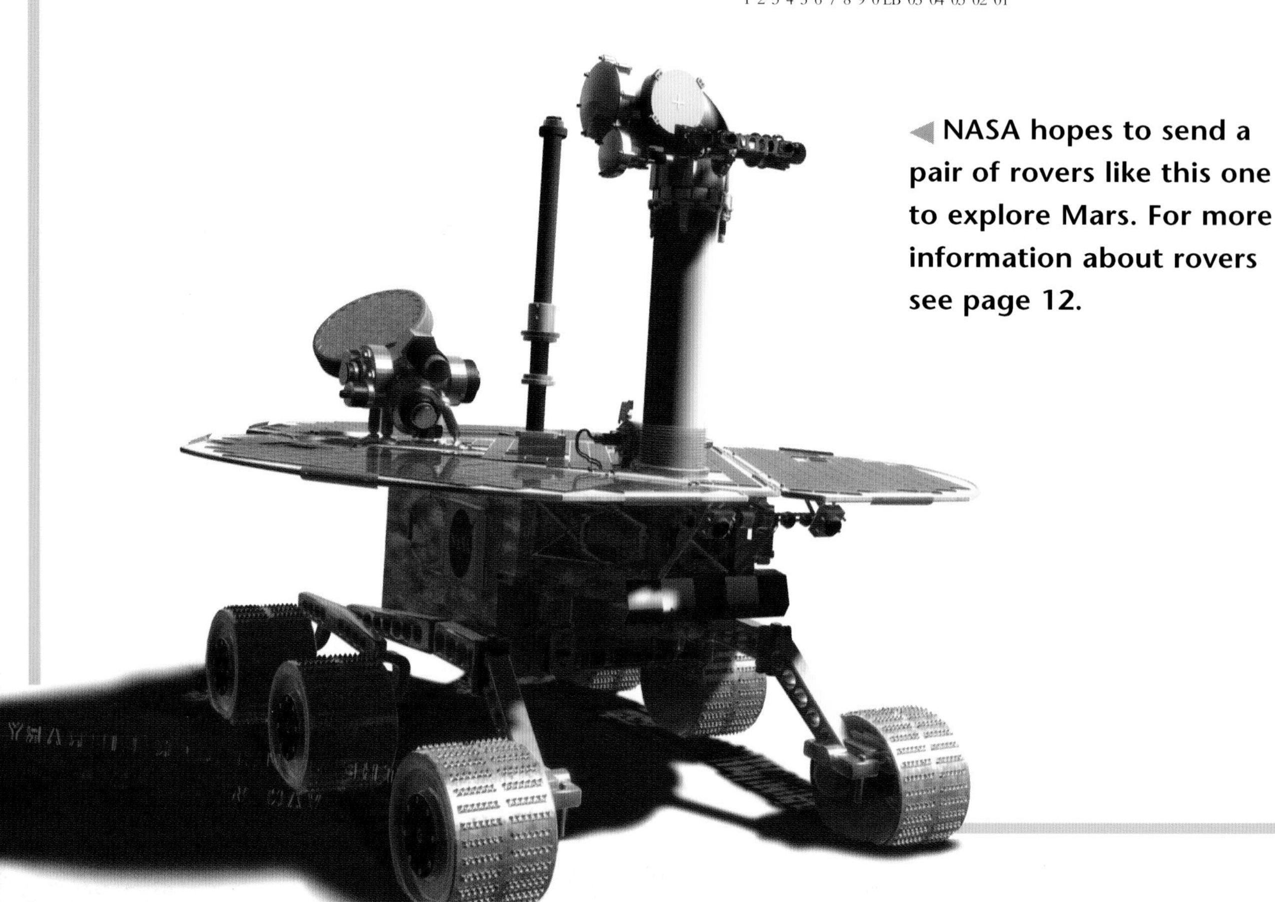

◀ **NASA hopes to send a pair of rovers like this one to explore Mars. For more information about rovers see page 12.**

Contents

Space Travel

Earth is one of nine planets that travel around a star we call the Sun. There are millions of other stars in our galaxy (group of stars). There are millions of other galaxies in space, too. Space is so huge, it is impossible to imagine.

◀ ***Gemini 10*** **blasting off in July 1966, three years before the first landing on the Moon.**

During the 1990s, American astronauts visited and worked on board the Russian *Mir* space station. Here a space shuttle has just docked (joined together) with *Mir*.

In the 1950s, the age of space travel began. The world's two strongest countries, the U.S. and the Soviet Union (a huge country made up of many small countries, including Russia), began to build rockets powerful enough to blast spacecraft into space.

In 1961, Yuri Gagarin of the Soviet Union became the first man in space. Eight years later, the American astronauts Neil Armstrong and Edwin "Buzz" Aldrin became the first people to land on the Moon, during the *Apollo 11* mission.

These were the first small steps in space travel. Today, astronauts live in space in floating buildings called space stations. There are also plans to send astronauts to Mars, the nearest planet to Earth.

THE FIRST ROCKETS

The first rocket with liquid fuel was invented by the American Robert Goddard in 1926. It only reached a height of 41 feet (12.5 m). But the idea behind it led scientists to build the huge Saturn rockets that blasted the *Eagle* lunar module on its journey to the Moon.

Robert Goddard, on the left, working on an early rocket engine.

PICTURES FROM PROBES

▲ The *Mariner 4* probe carried cameras to take pictures of the red surface of Mars.

Space probes are small spacecraft that do not need a pilot. They explore space all on their own! Probes take photographs of distant planets and their moons, and asteroids and comets. They also send information back to Earth.

In 1959 a probe launched by the Soviet Union, called *Luna 2*, landed on the Moon. It was the first human-made object to land somewhere other than Earth. There were 24 Luna probes in all. Later probes were sent to Mars. In 1965, an American probe called *Mariner 4* took the first close-up pictures of Mars.

CHASING COMETS

Probes can explore small, fast-moving objects, such as comets and asteroids. A European probe called *Giotto* flew close to Halley's comet in 1986. Giotto studied the core of the comet. In 2000, the United States' *NEAR* probe actually landed on an asteroid named Eros.

◀ The *Huygens* probe will land on Saturn's moon, Titan, in 2004. Some scientists think it might discover signs of life.

▼ The two Voyager probes explored Jupiter and Saturn.

The Venera probes from the Soviet Union were the first to send back pictures from the glowing-hot surface of Venus. More recently, an American probe called *Magellan* studied the volcanoes and craters of Venus.

Probes are traveling farther and farther into space. In 1995, a probe called *Galileo* began taking thousands of photographs of Jupiter. In 2004, a probe called *Cassini* will be ready to study Saturn.

Shooting Shuttles

The space shuttle is the first ever spacecraft that can travel into space again and again. It is a kind of space truck, which carries things into space and out again. It has carried telescopes, laboratories, and satellites. It has also taken probes into space, before launching them toward distant planets.

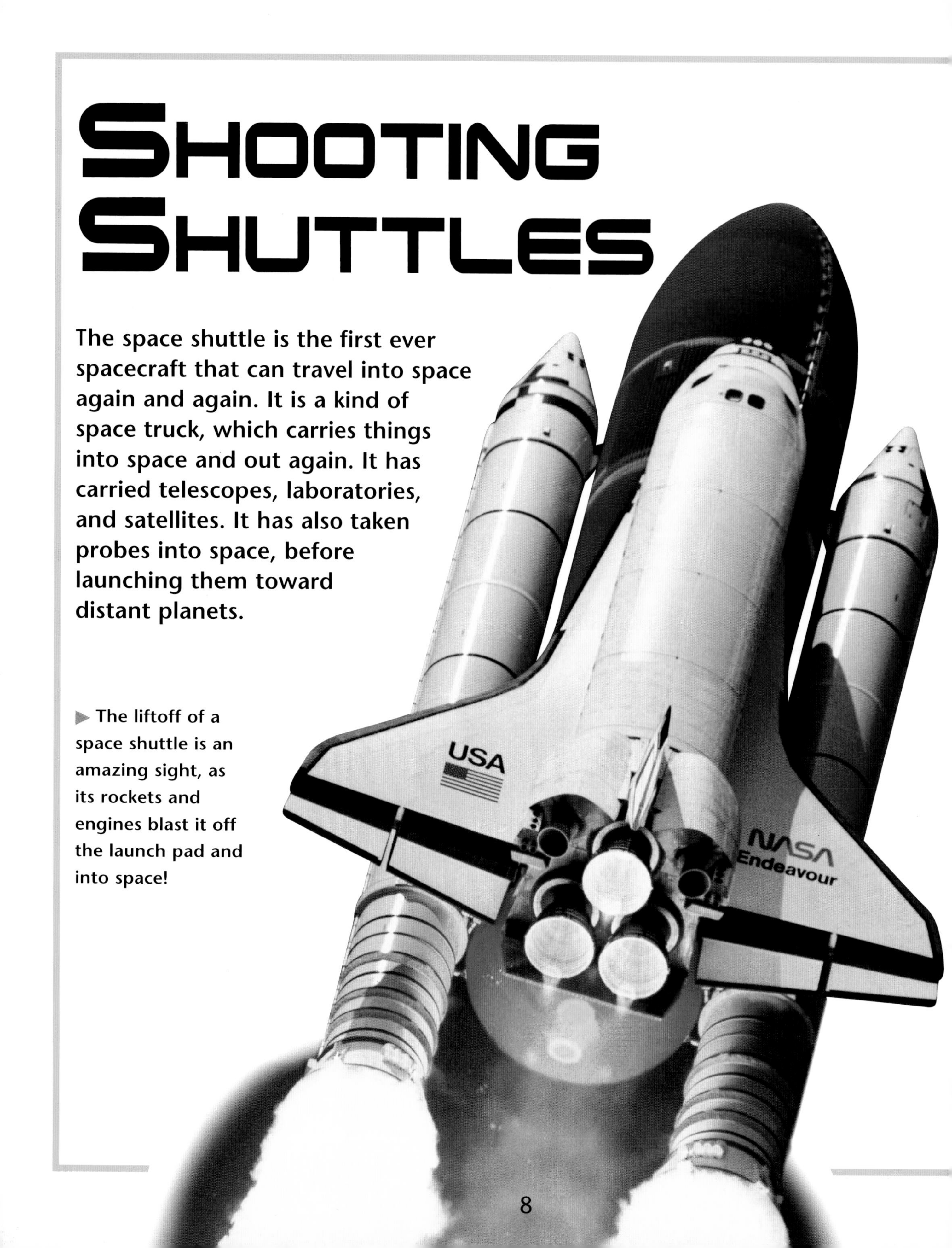

▶ The liftoff of a space shuttle is an amazing sight, as its rockets and engines blast it off the launch pad and into space!

▲ **Each orbiter has its own small kitchen, where astronauts prepare meals.**

Two powerful rockets lift the space shuttle off the launch pad and power it into space. The rockets burn for two minutes, then fall away and parachute into the ocean. A huge fuel tank falls away when empty, then burns up in the air, eight minutes after take-off.

This leaves a part called the "orbiter," which looks just like an airplane. The orbiter can circle Earth for up to two weeks, before landing on a runway.

Inside the orbiter, astronauts can wear just T-shirts and shorts. There is no gravity, so they can float from one place to another. During meals they have to hold on to their food to stop it from floating away!

PIGGY-BACK PLANE

If an orbiter cannot land back at the Kennedy Space Center, it has to use a runway built in a desert on the other side of the U.S. To get home, the orbiter has to hitch a "piggy-back" ride to Florida—on top of a special jumbo jet!

▼ **After it has landed at the Space Center, the orbiter is lifted off the back of the 747 jet with a special crane and towed back to its hangar (shed).**

TRAINING TRAVELERS

Both men and women can become astronauts. But it is very difficult. If you want to travel in space, you will have to work hard at school. You will also need to study math or science at a college or university.

To train as an astronaut in the U.S., you would join the Astronaut Corps at the Johnson Space Center in Texas. The astronauts are trained to fly fast jets, study science and math, and learn to use that knowledge in a special area.

▼ During a launch, shuttle crews wear pressure suits and space helmets to protect them in case anything goes wrong.

They would also learn all about gravity, by flying on a special aircraft. The KC-135 flies up and down like a roller coaster. As it drops suddenly, the people inside become weightless and float up off the floor, just like astronauts do in space. This makes some people sick—the plane is nicknamed Vomit Comet!

Astronauts learn how to fly a space shuttle by practicing on "flight simulators" that are like computer games. There is also a huge swimming pool. In the pool astronauts learn how difficult it is to walk when you are weightless in space.

▲ As the KC-135 drops, its passengers float up into the air.

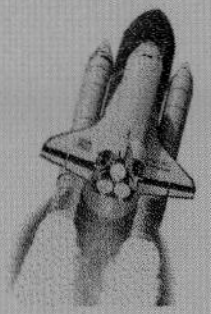

SPACE AGE

In 1962, John Glenn became the first American to orbit Earth. Thirty-six years later, he flew into space again on a space shuttle called *Discovery*. By then he was 77 years old.

▶ John Glenn's first spaceflight was inside a Mercury capsule called *Friendship 7*.

Robot Rovers

▲ A model of the *Lunokhod* rover that explored the Moon for almost a year.

Some space probes contain "robot rovers." After the probe lands, these rovers drive around to take a good look. The first rover was the Soviet Union's *Lunokhod 1*. It landed on the Moon in 1970 and slowly traveled around for over ten months.

In 1997, the American *Pathfinder* probe reached Mars. The probe was protected inside huge air bags. Parachutes carried it gently down, then *Pathfinder* bounced across the surface. Finally, it opened up, and a small rover called *Sojourner* drove out

Earth Rovers

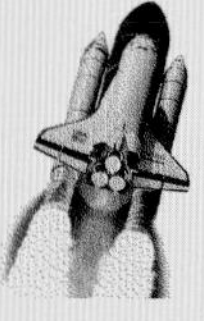

Rovers sent to explore other planets are controlled from Earth. Scientists steer them around the craters and rocks in their path by moving a joystick.

▶ **The little *Sojourner* rover studied many boulders on Mars, including this one nicknamed Yogi.**

▲ Future rovers sent to Mars will be larger and have better cameras and equipment.

Scientists guided the little rover from Earth, and *Sojourner* took amazing close-up pictures of rocks on Mars. The scientists gave the biggest rocks names, such as Yogi and Scooby Doo! *Sojourner* is still on Mars, waiting to be brought back to Earth.

In 2003, the U.S. may send two more rovers to Mars, to hunt for signs of water and life. Rovers may also be sent to the Moon to look for ice inside its craters. This ice could be used by astronauts in the future to make drinking water and even rocket fuel.

Space Station

▲ When the Space Station is finished, astronauts will live on it for up to six months at a time.

THINKING OF HOME

To stop astronauts from becoming too homesick, ISS crews will be able to keep in touch with their friends and family on Earth using radios, video links, and E-mail. They will probably have photographs of their family with them, too.

A huge spacecraft called the International Space Station (ISS) is being built in space at this very moment, over 236 miles (380 km) above you. Astronauts use it to learn how to live and survive in space for long periods of time.

The ISS is traveling around Earth in a loop, or "orbit." Several times a year, a space shuttle or rocket carries another part of the ISS up into space. This new part is then joined to the others. The ISS is like a huge model kit.

While they build the ISS, astronauts are more like builders than space explorers. They float in space, putting the parts together. Sometimes a powerful robot arm is used to swing pieces into place.

You can sometimes see the ISS from here on Earth. It has massive solar panels, which change sunlight into energy. When these panels reflect the sunlight, the ISS shines like a bright star drifting across the sky.

▼ **These spacewalking astronauts are working outside the ISS, high above Earth.**

SPACE PLACE

Astronauts travel to and from the ISS inside space shuttles. As one tired crew finishes its mission, another crew flies up to replace it.

Imagine working in space. Astronauts living on the ISS will perform experiments in special laboratories. They will also carry out medical tests, to see how their bodies are affected by being in space.

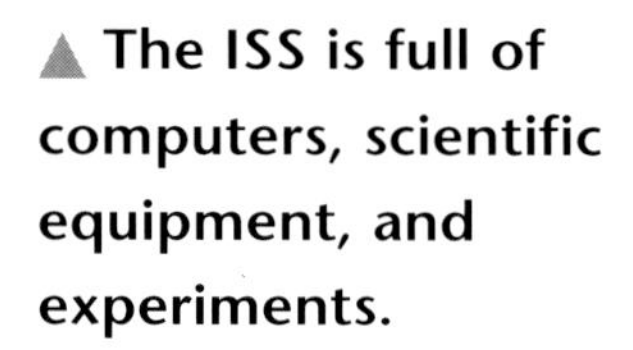

▲ The ISS is full of computers, scientific equipment, and experiments.

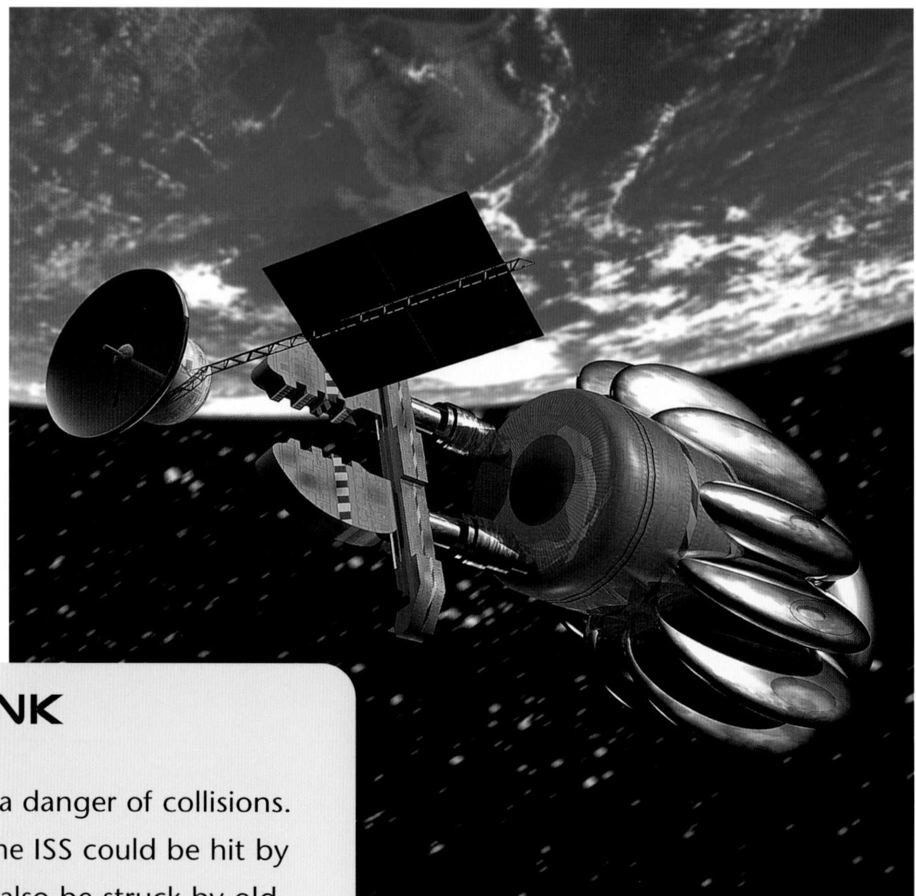

► In the future we may need to use special robots like this to collect pieces of orbiting space junk.

SPACE JUNK

Even in space there is a danger of collisions. Scientists worry that the ISS could be hit by fast-flying meteorites. It might also be struck by old satellites or spacecraft that are still drifting through space. These are known as "space junk."

But it will not be all work. The astronauts will be able to rest in private rooms, or get together to listen to music or watch films. They will probably spend a lot of time looking out of the windows. From the ISS, there are amazing views of Earth's beautiful oceans, forests, and mountains.

Until 2001, only professional astronauts had ever flown in space. That changed when a very rich American businessman, Dennis Tito, paid the Russian space agency $20 million to fly him to the ISS—as the very first space tourist. He stayed on board for a week before returning to Earth.

▶ **Dennis Tito said that being up in space, on board ISS, was "paradise."**

Moon Mining

The next astronauts who travel to the Moon will probably be staying. There are plans to build a Moon base underground. It will be big enough for two or three people to live in. As more people come to live on the Moon, the base will grow larger. Perhaps one day, children will be born on the Moon.

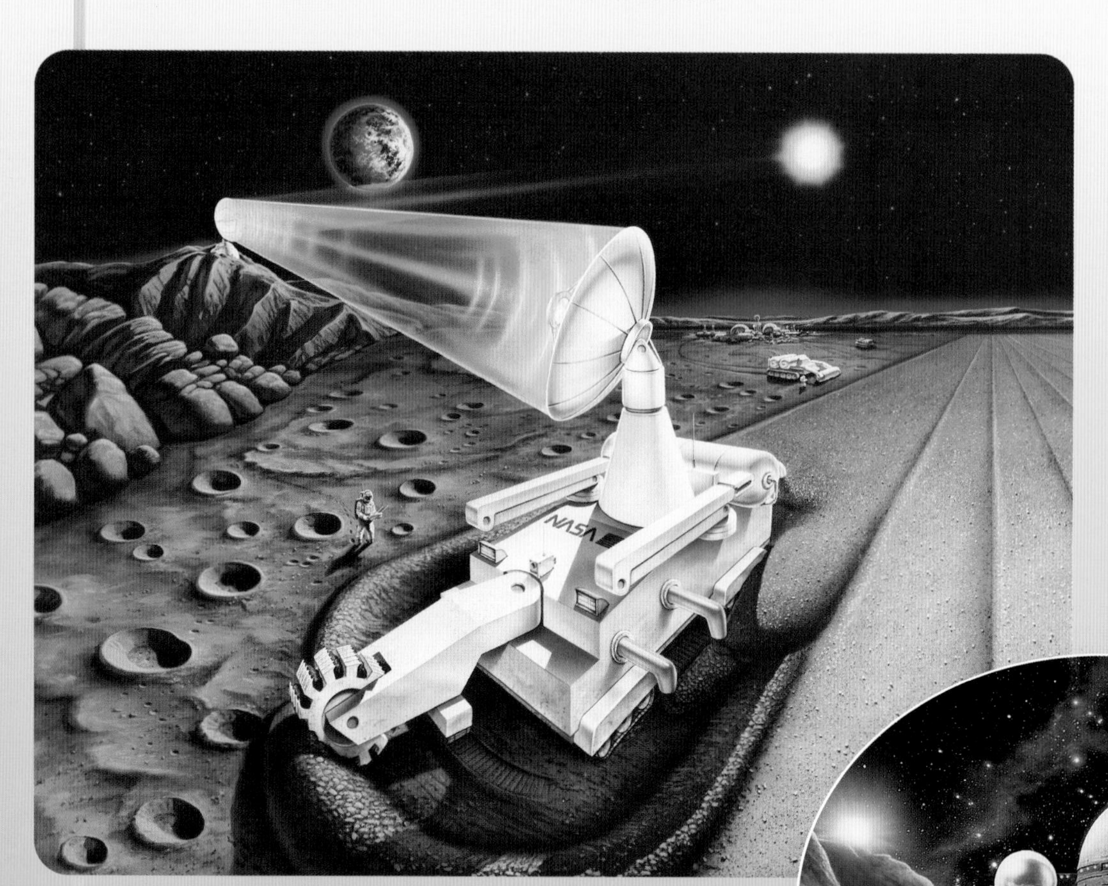

◀ In the future, solar-powered robot factories could mine minerals on the Moon.

▼ It is possible that one day towns and even cities will be built around the Moon's mines.

With its huge craters and seas of frozen lava, the Moon would make an exciting home. The Moon has no clouds, so astronomers could build giant telescopes and easily see and study distant planets.

Scientists could also test equipment and prepare for journeys to planets like Mars. People could dig mines on the Moon to look for valuable metals and minerals, too.

FOUR-STAR HOTEL

Several companies have plans to build "space hotels" for space travelers. These would be special space stations where people could take a break, enjoy the feeling of floating through space, and see amazing views of Earth from high above.

On Earth, we protect places where history was made. Maybe one day it will be the same on the Moon. The landing site of the *Apollo 11* mission could become a Moon museum. Tourists would be able to look at Neil Armstrong's first footprint in the dust.

▲ The first Moon base may be built in pieces, like the ISS. A large radio dish will let the people talk to Earth.

Martian Mission

Mars is known as the red planet. Scientists have studied it for hundreds of years. They have looked at it through telescopes and sent probes to its surface. Some people believe there may be water and very simple forms of life on Mars.

A piece of rock from Mars has been found in Antarctica. Some scientists say that it contains the remains of tiny Martian creatures. Others say these are just the remains of Earth creatures that crawled into the rock after it landed on planet Earth.

▲ Scientists are working toward the day when robot planes can fly low over Mars, through its canyons and valleys.

◀ Mars meteorite ALH84001 was found in Antarctica in 1984. Some scientists believe it contains very simple forms of life.

There are plans to send astronauts to Mars to look for signs of life. But Mars is millions of miles away, and a special spacecraft will be needed to fly the astronauts there. They will need excellent training. They will have to spend a long time on the ISS to get used to living in space. They might also train in Antarctica to practice living on Mars—where it is very, very cold, as low as –207° Fahrenheit (–133° degrees)!

MARTIAN CANALS

In 1877 an Italian astronomer named Giovanni Schiaparelli thought he saw straight lines on Mars through his telescope. Some people thought they were canals dug by Martians! Today, we know that they are just natural features.

◀ Olympus Mons on Mars is the biggest volcano in the Solar System. It is three times higher than Mount Everest.

Moving to Mars

A spacecraft big enough to travel to Mars would probably have to be built in space. It would have several parts, just like the ISS. With nuclear engines it might be able to reach Mars in six months.

Scientists have a plan for traveling to Mars. First, they would land a probe on the planet. This probe would be able to make fuel and water out of Mars' atmosphere. A spacecraft carrying astronauts would follow two years later. After landing on Mars, the astronauts would use the fuel made by the probe when they needed to fly home.

▼ **Scientists may be able to change Mars' red and rocky landscape into a place where humans could live, with fertile land and oceans.**

The first people to land on Mars might live there for a year, studying the planet and looking for life. Earth would look like a bright-blue star from so far away.

Some scientists think that one day we will build cities on Mars. They believe it might be possible to change the red planet so that it has forests, oceans, and a blue sky—just like Earth!

◀ **Some people even think that "the face" is a huge monument built by Egyptians who traveled to Mars.**

THE FACE ON MARS

An American probe called *Viking* took pictures of Mars in 1976 that showed what looked like a huge face on the surface. Some people thought it was a statue carved by aliens. But recent pictures show the face is a mountain, with hollows that look like eyes.

Looking for Life

Probes and spacecraft travel farther into space. As they do, we are learning more about our galaxy all the time. One day we might discover alien forms of life. They might be on a planet or on a moon that orbits a planet.

Europa is a moon that orbits Jupiter. Some astronomers think there might be simple forms of life beneath its icy surface. One day, Europa will be studied by probes and robot rovers.

▲ An astronaut standing on Europa would be surrounded by ice, and perhaps even simple life forms.

◀ Pictures of Europa taken by the *Galileo* probe have made many astronomers believe there is water beneath its surface.

MOVING MINES

Asteroids are enormous chunks of metal and rock that orbit the Sun. One day spacecraft might land on them and bring them back to Earth. It will be cheaper and cleaner to mine their metals and minerals than to dig them out of the ground on Earth!

Saturn's largest moon is called Titan. It is so big, it is more like a planet than a moon. Its surface is hidden beneath thick fog or smog that may hide forms of life. Some scientists believe it could be explored using an airship.

Films show space travelers going to distant stars, but this may never happen. Stars are so far away that even our fastest rockets would take many thousands of years to reach them. But who knows? Perhaps spacecraft of the future will travel to planets orbiting the suns of other galaxies.

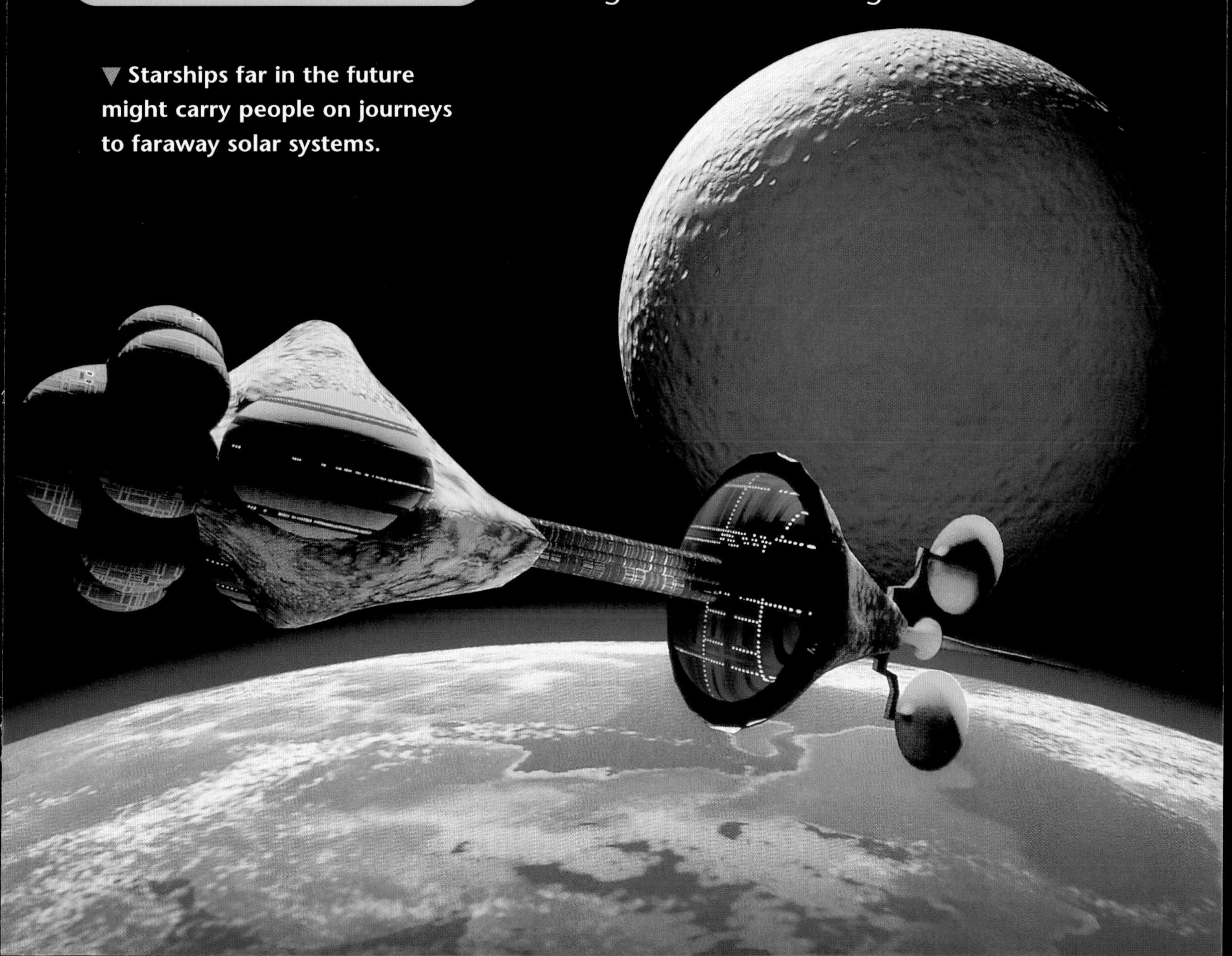

▼ Starships far in the future might carry people on journeys to faraway solar systems.

Space Travel Facts

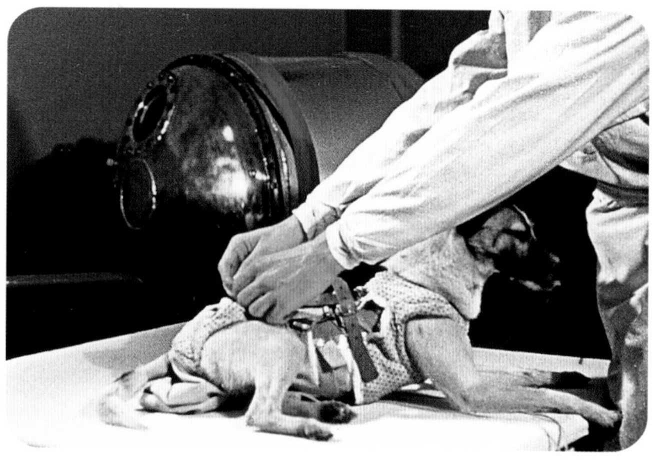

▲ Laika being prepared for her spaceflight.

Beep beep
The first human-made object put into space was a small satellite from the Soviet Union called *Sputnik*. It looked like a metal ball with four aerials sticking out of it and beeped as it traveled around Earth.

Dog star
The first living creature to be launched into space was a dog named Laika. She was a passenger on the Soviet Union's spacecraft *Sputnik 2* in 1957. Sadly, she did not survive the journey.

Space sounds
If any aliens find the U.S.'s Voyager probes in the future, they will find special gold-plated records on their sides. If the aliens work out how to play these disks, they will hear music, sounds of nature, and greetings from Earth's world leaders.

Big burn
When the *Galileo* probe finishes exploring Jupiter and its moons, it will burn up in the giant planet's atmosphere like a shooting star.

Moon walker
The first space shuttle was launched on April 12, 1981, and carried only two astronauts. One of them, John Young, had walked on the Moon nine years earlier on the *Apollo 16* mission.

Shifting shuttle
When the space shuttle is orbiting Earth, it is traveling at a speed of over 17,000 miles per hour (27,353 kmh). That is ten times faster than a bullet!

Heavy coat
On its first few flights, the space shuttle's huge fuel tank was painted white. But now it is left a rusty-orange color. The white paint weighed hundreds of pounds, and without it the shuttle can carry more into space.

Practice makes perfect

Astronauts are trained how to land the space shuttle by landing a plane that has been changed to make it fly like the shuttle. Before going on a shuttle mission, an astronaut must have completed 800 practice landings!

Price probe

The U.S.'s Mars *Pathfinder* probe and its *Sojourner* robot rover cost $196 million. But the cost of the Viking probes that landed on Mars 11 years earlier was nearly double that!

Crash course

The *Sojourner* robot rover was steered around the sharp rocks of Mars by scientists back on Earth. But a signal takes several minutes to reach Mars, so the rover's controllers had to drive it very slowly. Otherwise it would have crashed into rocks!

Super station

Sixteen different countries are working together to design and build the International Space Station. It will not be finished until around 2006. When the ISS is ready, it will have as much room inside as two jumbo jets!

Astronaut aerobics

While they are living and working on the ISS, astronauts must exercise every day using running or rowing machines. If they do not, their muscles and bones will soon become dangerously weak in space.

Sun run

The ISS orbits Earth once every 90 minutes. That means that if you were on board you would see 16 sunrises and sunsets every day!

Dust buster

When the *Mariner 9* space probe reached Mars in 1971, it could not see anything on the planet. The surface was hidden beneath a huge dust storm. When the dust cleared, *Mariner* discovered craters, towering volcanoes, and a huge valley.

Martian meteorites

Many pieces of rock from Mars have been found on Earth, where they arrived as meteorites. The most famous is called ALH84001. It was found in Antarctica in 1984 by Dr. Roberta Score. It may contain remains of Martians!

▲ **One day people may live on Mars in cities like this one.**

Space Travel Words

Galaxies are huge "star cities." Many bulge in the middle, like this one.

airship (AIR-ship)
A large balloon that carries people beneath it. Some scientists think planets could be explored using airships.

Antarctica (ant-ARK-tik-uh)
A huge frozen continent at Earth's South Pole. Astronauts and scientists may train there before going to Mars.

asteroid (AS-tuh-roid)
A large chunk of metal or rock that orbits the Sun like a tiny planet. Most asteroids lie in a group, called an asteroid belt, between the planets Mars and Jupiter.

astronaut (AS-truh-nawt)
Someone who travels into space.

astronomer (uh-STRON-uh-mur)
Someone who studies the stars and planets.

atmosphere (AT-muhss-fihr)
The layer of gases above the surface of a planet. Near to Earth's surface, the atmosphere is made up mainly of nitrogen. Higher up, there are other gases such as ozone.

flight simulators (flite SIM-yuh-lay-turz)
Machines in which astronauts can learn how to fly a spacecraft without leaving the ground. They are something like giant computer games.

galaxy (GAL-uhk-see)
A huge group of hundreds of millions of stars, all drifting through space together. Some galaxies are shaped like enormous spirals. Our own galaxy, which is called the Milky Way, is shaped like this.

fog
A thick, dense cloud.

gravity (GRAV-uh-tee)
Any large object has a natural force called gravity that pulls things toward it. The gravity of Earth is what holds us all on the ground and stops us from floating off into space! It also holds Earth in orbit around the Sun, and the Moon in orbit around Earth.

Halley's comet (HAYL-eez KOM-it)
Comets are huge chunks of dirty and dusty ice that travel around the Sun. When they melt, they grow glowing tails. The most famous one is Halley's comet, which travels past Earth every 76 years.

laboratory (LAB-ruh-tor-ee)
A place where scientists work and carry out experiments. Scientists can now work in special laboratories in space.

meteorite (MEE-tee-ru-rite)
A piece of space rock that falls to Earth after traveling around the Sun for many millions of years.

minerals (MIN-ur-uhlz)
Natural substances, such as rocks and crystals, which can be found in the ground on Earth and other planets.

moon
An object in space, like a small planet, that travels in orbit around a planet. Earth has one moon, but some other planets have more than one.

nuclear (NOO-klee-ur)
A type of power. When this fuel is used, huge amounts of energy are released.

orbit (OR-bit)
The path that a smaller object travels through space around a larger one. For example, the Moon travels in orbit around Earth. The Moon is held in this orbit by the pull of gravity from Earth. Spacecraft can also be sent into orbit around moons, stars, and planets.

planet (PLAN-it)
A very large object in space that travels around, or orbits, a star. It is usually round. The Sun is a star with nine planets orbiting it. These are, from the closest to the Sun to the farthest away: Mercury, Venus, Earth, Mars, Jupiter, Saturn, Uranus, Neptune, Pluto.

robot (ROH-bot)
A machine that can act like a human or do things a human can do.

rover (ROH-Vur)
A small probe with wheels, which can drive around a planet or moon for short distances after landing on its surface. It can be guided from Earth by the use of a radio control.

satellite (SAT-uh-lite)
A satellite is an object that flies through space, traveling around a large object such as Earth or the Moon in a big loop, or "orbit."

shooting star (SHOOT-ing)
The streak of light we see in the night sky. It is really a piece of rock flying through space that burns up in Earth's atmosphere.

solar panels (SOH-lur PAN-uhlz)
Long rows of tiles, which look like windows and can collect sunlight and turn it into electricity. Solar panels can be used to supply power to a spacecraft.

star
An enormous ball of very hot gases, which shines brightly in space. The Sun is a star.

weightless (WATE-liss)
Having no weight. Without gravity to pull them down, astronauts in space become weightless. They can float around freely.

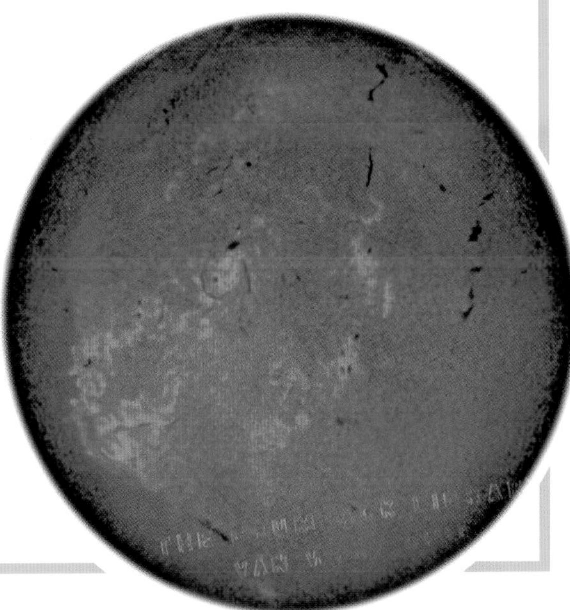

▶ The Sun is the nearest star to Earth.

Space Travel Projects

SPOT A SATELLITE

Go outside on the next clear night, about an hour after sunset, and look up at the sky. After a few minutes you will probably be able to see something that looks like a star moving slowly across the sky. Well done, you have just spotted a satellite! (If you see a blinking or flashing light, that is probably just a plane.) Sometimes you can see the International Space Station or even a space shuttle in the sky from where you live. You can find out how by looking at one of the websites listed on page 31.

ASTRONAUT REPORT

Visit your nearest park or beach, and search for interesting rocks. This is just what an astronaut would do on Mars. Imagine you are the first person to land on Earth. Write a report of what the planet looks like, and describe the different rocks you have found.

▶ **Hundreds of satellites are used to study the weather and to send television signals all around the world.**

PLAN A SHUTTLE FLIGHT

Imagine you are in charge of the next space shuttle flight from Earth to the International Space Station. Make a list of the different things you want to load on the shuttle to take up to the ISS. Use books and the websites listed below to help you.

SPACE SURFING

The Internet has so many great websites about space and space travel that it is impossible to visit them all. Some websites will tell you exactly when and where you can see the ISS. Have a look at some of the websites listed below. Try to find others by searching for keywords like "space shuttle" or "space probe."

▲ **A space shuttle on its launch pad lit by floodlights.**

NASA children's website
http://kids.msfc.nasa.gov
This is a children's website created by the American space agency that is called NASA.

Explore the Universe
Hubble Space Telescope: activities, resources.
http://www.stsci.edu

Greatest Hits of the Hubble 1990–1995.
http://www.opposite.stsci.edu/pubinfo/BestOfHST95.html

Pictures, captions, and press releases from the Hubble telescope.

Follow the Space Station
www.hq.nasa.gov/osf/station/viewing/issvis.html
This site will tell you where to see the ISS at different times.

Find out about astronaut training
www.jsc.nasa.gov
This is the website for the Johnson Space Center.

NASA's continuous broadcast
www.unitedspacealliance.com/live/NASATV.htm
Live pictures from space.

INDEX